# Math Mammoth
# Grade 1 Tests and
# Cumulative Reviews

for the complete curriculum
(International Version — Canada)

Includes consumable student copies of:

- Chapter Tests
- End-of-year Test
- Cumulative Reviews

*By Maria Miller*

# Contents

# Grade 1, Chapter 1

# End-of-Chapter Test

## Instructions to the student:

Answer each question in the space provided.

## Instructions to the teacher:

My suggestion for grading the chapter 1 test is below. The total is 28 points. Divide the student's score by the total of 28 to get a decimal number, and change that decimal to percent to get the student's percentage score.

| Question # | Max. points | Student score |
|---|---|---|
| 1 | 6 points | |
| 2 | 8 points | |
| 3 | 6 points | |

| Question # | Max. points | Student score |
|---|---|---|
| 4 | 4 points | |
| 5 | 4 points | |
| Total | 28 points | |

# Chapter 1 Test

1. Add.

a.
$$\begin{array}{r} 2 \\ +\ 4 \\ \hline \end{array}$$

b.
$$\begin{array}{r} 8 \\ +\ 1 \\ \hline \end{array}$$

c.
$$\begin{array}{r} 6 \\ +\ 2 \\ \hline \end{array}$$

d.
$$\begin{array}{r} 8 \\ +\ 2 \\ \hline \end{array}$$

e.
$$\begin{array}{r} 5 \\ +\ 4 \\ \hline \end{array}$$

f.
$$\begin{array}{r} 6 \\ +\ 3 \\ \hline \end{array}$$

2. Compare. Write $<$, $>$, or $=$.

| | | | |
|---|---|---|---|
| a.  5 ☐ 6 | c.  $6 + 1$ ☐ 10 | e.  3 ☐ $1 + 0$ | g.  8 ☐ $6 + 3$ |
| b.  2 ☐ 9 | d.  $2 + 3$ ☐ 5 | f.  10 ☐ $0 + 6$ | h.  7 ☐ $2 + 5$ |

3. Draw more. Write an addition sentence.

a. _____ + _____ = 10

b. _____ + _____ = 7

c. _____ + _____ = 10

4. Find the missing numbers.

a. $2 +$ _____ $= 7$   b. $1 +$ _____ $= 4$   c. $4 +$ _____ $= 10$   d. _____ $+ 7 = 9$

5. Solve the word problems.

a. Anna has seven stuffed animals and Abby has three. How many do they have in total?

b. Andy has eight pairs of shorts. Two of them are in the wash. How many are not?

# Grade 1, Chapter 2

# End-of-Chapter Test

## Instructions to the student:

Answer each question in the space provided.

## Instructions to the teacher:

My suggestion for grading the chapter 2 test is below. The total is 30 points. Divide the student's score by the total of 30 to get a decimal number, and change that decimal to percent to get the student's percentage score.

| Question # | Max. points | Student score |
|:---:|:---:|:---:|
| 1 | 4 points | |
| 2 | 4 points | |
| 3 | 6 points | |

| Question # | Max. points | Student score |
|:---:|:---:|:---:|
| 4 | 16 points | |
| Total | 30 points | |

# Chapter 2 Test

1. Write the fact family to match the picture.

_____ + _____ = _____          _____ + _____ = _____

_____ − _____ = _____          _____ − _____ = _____

$$\boxed{\begin{array}{c} 8 \\ \circ\circ \; / \; \substack{\bullet\bullet\bullet \\ \bullet\bullet\bullet} \end{array}}$$

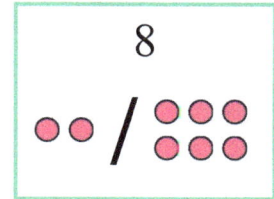

2. **a.** Write a subtraction sentence that matches the
   addition $5 + 4 = 9$, using the same numbers.          _____ − _____ = _____

   **b.** Solve the addition $6 +$ _____ $= 10$ and
   write a subtraction that matches the addition.          _____ − _____ = _____

3. **a.** There are 9 animals playing in the yard. Three are
   dogs and the rest are cats. How many cats are there?

   **b.** Erika has four more balls than Karen.
   Karen has five balls.
   Draw Karen's and Erika's balls.

   **c.** Five starlings and two swallows are feeding on seeds.
   Two more starlings fly in. Now how many more
   starlings are there than swallows?

4. Find the missing numbers.

| a. $4 +$ _____ $= 6$ | b. $9 -$ _____ $= 3$ | c. $9 - 0 =$ _____ | d. $6 - 5 =$ _____ |
|---|---|---|---|
| $1 +$ _____ $= 8$ | $7 -$ _____ $= 5$ | $7 - 1 =$ _____ | $3 - 2 =$ _____ |
| $5 +$ _____ $= 10$ | _____ $- 1 = 6$ | $9 - 7 =$ _____ | $4 - 4 =$ _____ |
| $6 +$ _____ $= 9$ | _____ $- 2 = 4$ | $8 - 2 =$ _____ | $10 - 7 =$ _____ |

# Grade 1, Chapter 3

# End-of-Chapter Test

### Instructions to the student:

Answer each question in the space provided.

### Instructions to the teacher:

My suggestion for grading the chapter 3 test is below. The total is 27 points. Divide the student's score by the total of 27 to get a decimal number, and change that decimal to percent to get the student's percentage score.

| Question # | Max. points | Student score |
|------------|-------------|---------------|
| 1 | 4 points | |
| 2 | 5 points | |
| 3 | 5 points | |
| 4 | 4 points | |

| Question # | Max. points | Student score |
|------------|-------------|---------------|
| 5 | 6 points | |
| 6 | 3 points | |
| Total | 27 points | |

# Chapter 3 Test

1. Name the numbers (with words).

   **a.** 1 ten 6 ones _____

   **b.** 7 tens 8 ones _____

   **c.** 5 tens 1 ones _____

   **d.** 9 tens 0 ones _____

2. Fill in the missing numbers on the number line.

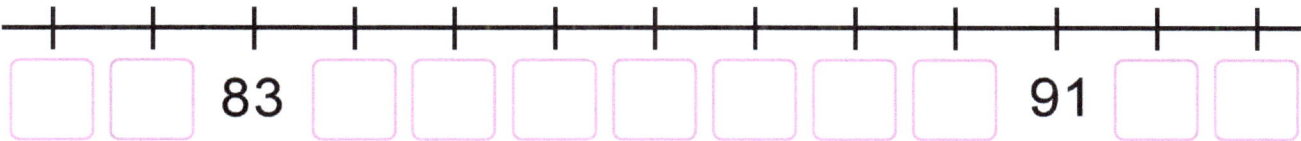

   ☐ ☐ **83** ☐ ☐ ☐ ☐ ☐ ☐ **91** ☐ ☐

3. Break the numbers into tens and ones.

   | | | |
   |---|---|---|
   | **a.** $45 = 40 + 5$ | **b.** $52 = $ _____ $ + $ ____ | **c.** $97 = $ _____ $ + $ ____ |
   | $86 = $ _____ $ + $ ____ | $32 = $ _____ $ + $ ____ | $19 = $ _____ $ + $ ____ |

4. Add the tens and ones.

   | **a.** | **b.** | **c.** | **d.** |
   |---|---|---|---|
   | $20 + 9 = $ _____ | $5 + 70 = $ _____ | $2 + 80 = $ _____ | $1 + 90 = $ _____ |

5. Put the numbers in order from the smallest to the greatest.

   | **a.** 75, 71, 57 | **b.** 69, 98, 96 | **c.** 81, 84, 49 |
   |---|---|---|
   | _____ < _____ < _____ | _____ < _____ < _____ | _____ < _____ < _____ |

6. Compare and write $<$, $>$ or $=$.

   **a.** 65 ☐ $5 + 60$   **b.** 43 ☐ $60 + 4$   **c.** $90 + 3$ ☐ $30 + 9$

# Grade 1, Chapter 4

# End-of-Chapter Test

### Instructions to the student:

Answer each question in the space provided.

### Instructions to the teacher:

My suggestion for grading the chapter 4 test is below. The total is 32 points. Divide the student's score by the total of 32 to get a decimal number, and change that decimal to percent to get the student's percentage score.

| Question # | Max. points | Student score |
|:---:|:---:|:---:|
| 1 | 14 points | |
| 2 | 6 points | |

| Question # | Max. points | Student score |
|:---:|:---:|:---:|
| 3 | 12 points | |
| Total | 32 points | |

# Chapter 4 Test

1. Find the missing numbers.

**a.** $2 + \underline{\quad} = 8$

$3 + \underline{\quad} = 8$

$1 + \underline{\quad} = 8$

$4 + \underline{\quad} = 8$

**b.** $2 + \underline{\quad} = 9$

$3 + \underline{\quad} = 9$

$5 + \underline{\quad} = 9$

$1 + \underline{\quad} = 9$

**c.** $2 + \underline{\quad} = 10$

$3 + \underline{\quad} = 10$

$6 + \underline{\quad} = 10$

$1 + \underline{\quad} = 10$

**d.**

$8 - 5 = \underline{\quad}$

$8 - 7 = \underline{\quad}$

$8 - 2 = \underline{\quad}$

$8 - 3 = \underline{\quad}$

**e.**

$9 - 8 = \underline{\quad}$

$9 - 9 = \underline{\quad}$

$9 - 7 = \underline{\quad}$

$9 - 4 = \underline{\quad}$

**f.**

$10 - 4 = \underline{\quad}$

$10 - 9 = \underline{\quad}$

$10 - 7 = \underline{\quad}$

$10 - 8 = \underline{\quad}$

**g.**

$7 - 3 = \underline{\quad}$

$6 - 5 = \underline{\quad}$

$7 - 6 = \underline{\quad}$

$6 - 3 = \underline{\quad}$

2. Solve.

**a.** Sally has seven coins. Liz has three coins. Today, Liz found five more coins.
   Now who has more coins?
   How many more?

**b.** Dan had two boxes of nails. Then he bought four more boxes of nails.
   The next day he gave three boxes to the neighbour.
   How many boxes of nails does Dan have now?

3. **a.** Complete. Then connect with a line the facts from the same fact family.

$\underline{\quad} - 4 = 3$

$3 + \underline{\quad} = 5$

$8 - \underline{\quad} = 5$

$8 - 3 = \underline{\quad}$

$\underline{\quad} + 3 = 7$

$5 - 2 = \underline{\quad}$

   **b.** Complete. Then connect with a line the facts from the same fact family.

$2 + \underline{\quad} = 6$

$7 - \underline{\quad} = 4$

$\underline{\quad} + 3 = 9$

$\underline{\quad} - 4 = 3$

$\underline{\quad} - 6 = 3$

$2 + 4 = \underline{\quad}$

# Grade 1, Chapter 5

# End-of-Chapter Test

### *Instructions to the student:*

Answer each question in the space provided.

### *Instructions to the teacher:*

My suggestion for grading the chapter 5 test is below. The total is 34 points. Divide the student's score by the total of 34 to get a decimal number, and change that decimal to percent to get the student's percentage score.

| Question # | Max. points | Student score |
|:---:|:---:|:---:|
| 1 | 8 points | |
| 2 | 12 points | |
| 3 | 10 points | |

| Question # | Max. points | Student score |
|:---:|:---:|:---:|
| 4 | 4 points | |
| Total | 34 points | |

# Chapter 5 Test

1. Write the time using the expressions *o'clock* and *half past*.

a. _____ 

_____

b. _____

_____

c. _____

_____

d. _____

_____

2. Write the time in two ways: using *o'clock* or *half past*, and with numbers.

a._____

_____

_____ : _____

b._____

_____

_____ : _____

c._____

_____

_____ : _____

d._____

_____

_____ : _____

3. Write the time for a half-hour and an hour later from the given time. Use numbers.

| Now it is: | a. 6:00 | b. 9:30 | c. 10:00 | d. 4:30 | e. 12:30 |
|---|---|---|---|---|---|
| a half-hour later, it is: | | | | | |
| an hour later, it is: | | | | | |

4. Fill in either AM or PM.

a. Anna wakes up. It is 7 _____.

b. Anna plays in the afternoon. It is 3 _____.

c. Anna sleeps. It is dark. It is 3 _____.

d. Time for an evening snack!  It is 7 _____.

## Grade 1, Chapter 6

# End-of-Chapter Test

**Instructions to the student:**

Answer each question in the space provided.

**Instructions to the teacher:**

My suggestion for grading the chapter 6 test is below. The total is 9 points. Divide the student's score by the total of 9 to get a decimal number, and change that decimal to percent to get the student's percentage score.

| Question # | Max. points | Student score |
|:---:|:---:|:---:|
| 1 | 2 points | |
| 2 | 2 points | |
| 3 | 3 points | |

| Question # | Max. points | Student score |
|:---:|:---:|:---:|
| 4 | 2 points | |
| Total | 9 points | |

# Chapter 6 Test

1. The two shapes are put together.
   What new shape is formed?

   a. _____

   b. _____

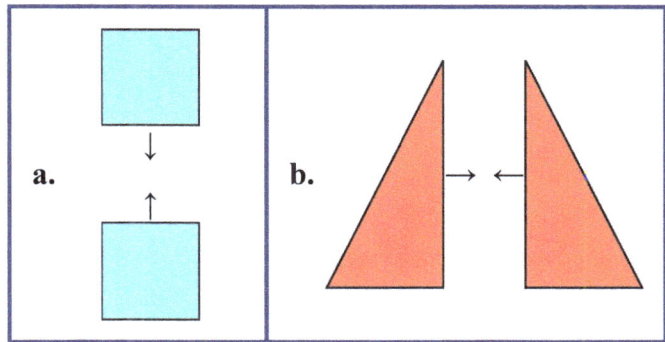

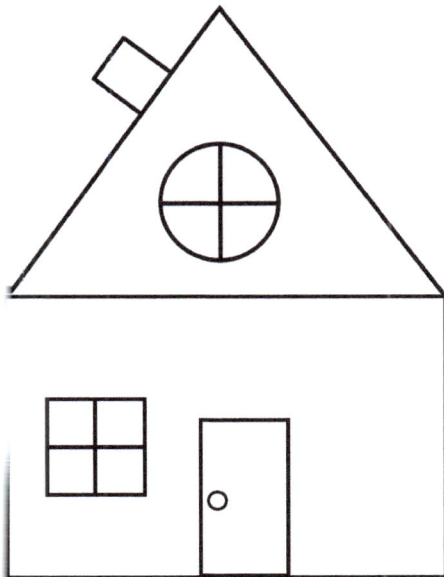

2. Colour the triangle red, one rectangle
   green and the other purple,
   the squares yellow, and
   the circles light blue.

3. Join these dots carefully with lines.
   Use a ruler. What shape do you get?

   _____

   Measure the sides of your shape
   in centimetres.

4. Draw a line that is:

   a. 10 centimetres

   b. 12 centimetres

# Grade 1, Chapter 7

# End-of-Chapter Test

***Instructions to the student:***

Answer each question in the space provided.

***Instructions to the teacher:***

My suggestion for grading the chapter 7 test is below. The total is 25 points. Divide the student's score by the total of 25 to get a decimal number, and change that decimal to percent to get the student's percentage score.

| Question # | Max. points | Student score |
|:---:|:---:|:---:|
| 1 | 3 points | |
| 2 | 6 points | |
| 3 | 2 points | |

| Question # | Max. points | Student score |
|:---:|:---:|:---:|
| 4 | 8 points | |
| 5 | 6 points | |
| Total | 25 points | |

# Chapter 7 Test

1. Add and subtract.

| a. $22 - 4 = $ _____ | b. $40 + 30 = $ _____ | c. $90 - 20 = $ _____ |
|---|---|---|
| $41 + 5 = $ _____ | $76 + 10 = $ _____ | $80 - 70 = $ _____ |

2. Add. First, make a new ten with some of the little dots.

| a. $25 - 38 = $ _____ | b. $14 + 25 = $ _____ | c. $27 + 27 = $ _____ |
|---|---|---|

3. Add. You can use the trick with nine and the trick with eight.

| a. | b. | c. | d. |
|---|---|---|---|
| $9 + 9 = $ _____ | $4 + 9 = $ _____ | $8 + 5 = $ _____ | $8 + 7 = $ _____ |

4. Add and subtract. Write the numbers under each other.

    a. $20 + 57$      b. $78 - 44$      c. $45 + 13$      d. $87 - 20$

5. Jake had 6 dollars and Jim had 12. Then, Jake got 10 dollars more.

   Now who has more money?

   How many dollars more?

# Grade 1, Chapter 8

# End-of-Chapter Test

**Instructions to the student:**

Answer each question in the space provided.

**Instructions to the teacher:**

My suggestion for grading the chapter 8 test is below. The total is 13 points. Divide the student's score by the total of 13 to get a decimal number, and change that decimal to percent to get the student's percentage score.

| Question # | Max. points | Student score |
|---|---|---|
| 1 | 6 points | |
| 2 | 3 points | |

| Question # | Max. points | Student score |
|---|---|---|
| 3 | 4 points | |
| Total | 13 points | |

# Chapter 8 Test

1. How much money? Write the amount in cents.

| | | |
|---|---|---|
| **a.**    _____ ¢ | **b.**    _____ ¢ | **c.**    _____ ¢ |
| **d.**    _____ ¢ | **e.**    _____ ¢ | **f.**    _____ ¢ |

2. Draw to make these amounts of money.

| | | |
|---|---|---|
| **a.** 65¢ | **b.** 90¢ | **c.** 45¢ |

3. You buy an item. How much money will you have left?

| | |
|---|---|
| **a. You have:**    If you buy a comb for 60¢, how much money is left?    _____ ¢ | **b. You have:**    If you buy a toy for 75¢, how much money is left?    _____ ¢ |

# Math Mammoth End-of-Year Test - Grade 1
## International Version (Canada)

This test is quite long, so I do not recommend that you have your child/student do it in one sitting. Break it into parts and administer them either on consecutive days, or perhaps on morning/evening/morning. Use your judgement.

This is to be used as a diagnostic test. Thus, you may even skip those areas and concepts that you already know for sure your student has mastered.

The test does not cover every single concept that is covered in *Math Mammoth Grade 1*, but all of the major concepts and ideas are tested here. This test is evaluating the child's ability in the following content areas:

- basic addition and subtraction facts within 0-10
- two-digit numbers
- adding and subtracting two-digit numbers
- basic word problems
- clock to the nearest half hour
- measuring and geometry (shapes)
- counting coins

**Note 1:** If the child cannot read, the teacher can read the questions.

**Note 2:** Problems #1 and #2 are done <u>orally and timed</u>. Let the student see the problems. Read each problem aloud, and wait a maximum of 5 seconds for an answer. Mark the problem as right or wrong according to the student's (oral) answer. Mark it wrong if there is no answer. Then you can move on to the next problem.

You do not have to mention to the student that the problems are timed or that he/she will have 5 seconds per answer, because the idea here is not to create extra pressure by the fact it is timed, but simply to check if the student has the facts memorized (quick recall). You can say for example (vary as needed):

*"I will ask you some addition and subtraction questions. Try to answer them as quickly as possible. In each question, I will only wait a little while for you to answer, and if you don't say anything, I will move on to the next problem. So just try your best to answer the questions as quickly as you can."*

In order to continue with the Math Mammoth Grade 2, I recommend that the child gain a minimum score of 80% on this test, and that the teacher or parent review with him any content areas that are found weak. Children scoring between 70 and 80% may also continue with grade 2, depending on the types of errors (careless errors or not remembering something, vs. lack of understanding). Again, use your judgement.

*Instructions to the student:*

Answer each question in the space provided.

*Instructions to the teacher:*

My suggestion for grading is below. The total is 104 points. A score of 83 points is 80%. A score of 73 points is 70%.

| Question | Max. points | Student score |
|---|---|---|
| **Basic Addition and Subtraction Facts within 0-10** | | |
| 1 | 8 points | |
| 2 | 8 points | |
| 3 | 4 points | |
| 4 | 8 points | |
| | *subtotal* | / 28 |
| **Place Value and Two-Digit Numbers** | | |
| 5 | 6 points | |
| 6 | 4 points | |
| 7 | 3 points | |
| | *subtotal* | / 13 |
| **Adding and Subtracting Two-Digit Numbers** | | |
| 8 | 6 points | |
| 9 | 6 points | |
| 10 | 4 points | |
| 11 | 3 points | |
| | *subtotal* | / 19 |

| Question | Max. points | Student score |
|---|---|---|
| **Basic Word Problems** | | |
| 12 | 2 points | |
| 13 | 2 points | |
| 14 | 2 points | |
| 15 | 2 points | |
| 16 | 2 points | |
| 17 | 6 points | |
| 18 | 6 points | |
| | *subtotal* | / 22 |
| **Clock** | | |
| 19 | 6 points | |
| 20 | 4 points | |
| | *subtotal* | / 10 |
| **Geometry and Measuring** | | |
| 21 | 2 points | |
| 22 | 5 points | |
| | *subtotal* | / 7 |
| **Money** | | |
| 23 | 3 points | |
| 24 | 2 points | |
| | *subtotal* | / 5 |
| | | |
| **TOTAL** | | **/ 104** |

# End-of-Year Test - Grade 1

## Basic Addition and Subtraction Facts within 0-10

In problems 1 and 2, your teacher will read you the addition and subtraction questions. Try to answer them as quickly as possible. In each question, he/she will only wait a little while for you to answer, and if you don't say anything, your teacher will move on to the next problem. So, just try your best to answer the questions as quickly as you can.

1. Add.

|  a. |  b. |  c. |  d. |
|---|---|---|---|
| $2 + 3 =$ _____ | $7 + 3 =$ _____ | $6 + 2 =$ _____ | $5 + 5 =$ _____ |
| $4 + 4 =$ _____ | $5 + 4 =$ _____ | $4 + 6 =$ _____ | $2 + 4 =$ _____ |
| $1 + 6 =$ _____ | $3 + 6 =$ _____ | $2 + 5 =$ _____ | $9 + 1 =$ _____ |
| $2 + 7 =$ _____ | $1 + 7 =$ _____ | $6 + 2 =$ _____ | $5 + 3 =$ _____ |

2. Subtract.

|  a. |  b. |  c. |  d. |
|---|---|---|---|
| $8 - 3 =$ _____ | $5 - 3 =$ _____ | $7 - 3 =$ _____ | $10 - 3 =$ _____ |
| $6 - 4 =$ _____ | $7 - 4 =$ _____ | $9 - 4 =$ _____ | $5 - 4 =$ _____ |
| $10 - 6 =$ _____ | $9 - 6 =$ _____ | $4 - 3 =$ _____ | $8 - 6 =$ _____ |
| $8 - 7 =$ _____ | $6 - 3 =$ _____ | $10 - 7 =$ _____ | $9 - 7 =$ _____ |

3. Write a fact family to match the picture.

_____ + _____ = _____       _____ + _____ = _____

_____ − _____ = _____       _____ − _____ = _____

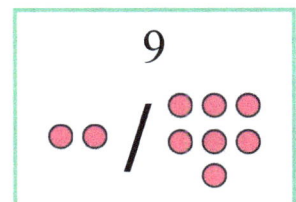

4. Find the missing numbers.

| a. $2 + \underline{\hspace{1cm}} = 7$ | b. $1 + \underline{\hspace{1cm}} = 8$ | c. $4 + \underline{\hspace{1cm}} = 6$ | d. $\underline{\hspace{1cm}} + 3 = 8$ |
|---|---|---|---|
| $3 + \underline{\hspace{1cm}} = 8$ | $2 + \underline{\hspace{1cm}} = 10$ | $\underline{\hspace{1cm}} + 3 = 9$ | $\underline{\hspace{1cm}} + 6 = 10$ |

## Place Value and Two-Digit Numbers

5. Fill in the missing parts.

| a. $20 + 7 = \underline{\hspace{1cm}}$ | b. $6 + \underline{\hspace{1cm}} = 56$ | c. $40 + \underline{\hspace{1cm}} = 40$ |
|---|---|---|
| $5 + 60 = \underline{\hspace{1cm}}$ | $30 + \underline{\hspace{1cm}} = 39$ | $4 + \underline{\hspace{1cm}} = 94$ |

6. Put the numbers in order.

| a. 16, 61, 26 | b. 54, 14, 51 |
|---|---|
| $\underline{\hspace{1cm}} < \underline{\hspace{1cm}} < \underline{\hspace{1cm}}$ | $\underline{\hspace{1cm}} < \underline{\hspace{1cm}} < \underline{\hspace{1cm}}$ |

7. Compare the expressions and write $<$, $>$, or $=$.

   a. $40 + 8$ [ ] $4 + 80$     b. $43 + 5$ [ ] $50$     c. $3 + 33$ [ ] $36$

## Adding and Subtracting Two-Digit Numbers

8. Add.

| a. $84 + 4 = \underline{\hspace{1cm}}$ | b. $6 + 70 = \underline{\hspace{1cm}}$ | c. $74 + 5 = \underline{\hspace{1cm}}$ |
|---|---|---|
| $41 + 4 = \underline{\hspace{1cm}}$ | $16 + 2 = \underline{\hspace{1cm}}$ | $6 + 53 = \underline{\hspace{1cm}}$ |

9. Subtract.

| a. $80 - 30 = \underline{\hspace{1cm}}$ | b. $55 - 3 = \underline{\hspace{1cm}}$ | c. $29 - 3 = \underline{\hspace{1cm}}$ |
|---|---|---|
| $17 - 3 = \underline{\hspace{1cm}}$ | $100 - 40 = \underline{\hspace{1cm}}$ | $50 - 2 = \underline{\hspace{1cm}}$ |

10. Add and subtract.

a.
$$14 + 35$$

b.
$$59 - 34$$

c.
$$40 + 56$$

d.
$$96 - 60$$

11. Add. The images can help you.

a. 19 + 34 = _____

b. 25 + 25 = _____

c. 22 + 27 = _____

## Basic Word Problems

12. Write a subtraction sentence that matches with the addition 6 + 8 = 14.

_____ − _____ = _____

13. How many more is 70 than 50?     _____ more

14. Henry owns four more cars than Mark, and Mark owns six cars.
    Draw Mark's cars and Henry's cars.

15. Ten children are playing in the yard. There are 6 boys. How many girls are there?

16. Andrew had 20 dollars. He bought a book for 10 dollars and another for 5 dollars.
    How much money does he have left?

17. A parking lot has 30 spaces for cars. There is a car in 22 of those spaces.

   **a.** How many spaces are empty?

   **b.** Now, two more cars drive in. How many cars are now in the parking lot?

   **c.** How many empty spaces are there now?

18. Isabelle had 70 marbles and her sister had 55. Isabelle gave 10 marbles to her sister.

   **a.** Now how many marbles does Isabelle have?

   **b.** And her sister?

   **c.** Who has more? How many more?

## Clock

19. Write the time in two ways: using *o'clock* or *half past*, and with numbers.

**a.** _____

_____

_____ : _____

**b.** _____

_____

_____ : _____

**c.** _____

_____

_____ : _____

20. Write the time for a half-hour and an hour later from the given time. Use numbers.

| Now it is: | a. 5:30 | b. 12:00 |
|---|---|---|
| a half-hour later, it is: | | |
| an hour later, it is: | | |

21. Draw a line that is:

    **a.** 3 centimetres

    **b.** 9 centimetres

22. **a.** Join these dots carefully with a ruler so that you get a shape.

    **A .**                                          **. B**




    **D ˙**                                          **˙ C**

    **b.** What is this shape called? _____

    **c.** Measure the sides of your shape in centimetres.

       Side AB: _____ cm        Side BC: _____ cm

    **d.** Draw a straight line from dot A to dot C. The line divides your shape to two
       new shapes.

       What are the new shapes called? _____

23. How much money? Write the amount in cents.

| | | |
|---|---|---|
| a.  _____¢ | b.  _____¢ | c.  _____¢ |

24. Solve.

You have:

You bought an apple for 55¢.

How much money do you have left? _____¢

# Using the cumulative reviews and the worksheet maker

The student books contain mixed review lessons which revise concepts from earlier chapters. The curriculum also comes with additional cumulative review lessons, which are just like the mixed review lessons in the student books, with a mix of problems covering various topics.

These cumulative reviews are optional; use them as needed. They are named indicating which chapters of the main curriculum the problems in the review come from. For example, the review titled "Cumulative Review, Chapters 1 - 4" includes problems that cover topics from chapters 1-4.

In the digital version of the curriculum, the cumulative reviews are provided both as PDF files and as html files. Again, the html versions are editable.

Both the mixed and cumulative reviews allow you to spot areas that the student has not grasped well or has forgotten. Another sign that the student has not understood a concept or skill is if he/she cannot do word problems in the curriculum that require that concept or skill.

When you find such a topic or concept, you have several options:

1. Check if the <u>worksheet maker</u> lets you make worksheets for that topic (for example, conversions between measuring units or equivalent fractions).

2. Check for any <u>online games and resources</u> in the Introduction part of the particular chapter in which this topic or concept was taught.

3. If you have the digital version, you could simply <u>reprint the lesson</u> from the student worktext, and have the student restudy that.

4. Perhaps you only assigned 1/2 or 2/3 of the exercise sets in the student book at first, and can now <u>use the remaining exercises</u>.

5. Check if our online practice area at https://www.mathmammoth.com/practice/ has something for that topic. We are constantly adding more exercises and games to this.

6. Khan Academy has free online exercises, articles, and videos for most any math topic imaginable.

# Cumulative Review, Grade 1, Chapters 1-2

1. Draw arrows to show the additions and subtractions.

**a.** 6 + 2 = _____

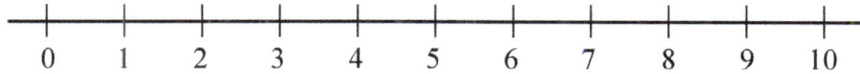

**b.** 9 − 4 = _____

**c.** 7 − 3 = _____

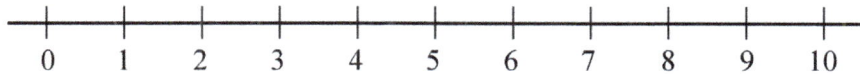

2. Make an addition sentence and a subtraction sentence from the same picture.

**a.**

_____ + _____ = _____

_____ − _____ = _____

**b.**

_____ + _____ = _____

_____ − _____ = _____

3. Add. Remember, you can add in any order.

| a. | b. | c. | d. | e. |
|---|---|---|---|---|
| 2 | 5 | 2 | 6 | 4 |
| 1 | 1 | 1 | 1 | 0 |
| + 4 | + 4 | + 1 | + 2 | + 3 |

4. Fill in the missing numbers.

| | | |
|---|---|---|
| a. $\triangle + 0 = 7$ | b. $\triangle - 2 = 4$ | c. $8 - \triangle = 1$ |

5. Compare. Write < , > or = .

| | | |
|---|---|---|
| a. $2 + 3 \square 5$ | c. $7 - 1 \square 9$ | e. $2 \square 4 - 2$ |
| b. $6 + 1 \square 8$ | d. $4 - 4 \square 0$ | f. $9 \square 9 - 1$ |

6. Find the missing numbers.

| a. | b. | c. | d. |
|---|---|---|---|
| $6 + 4 = \underline{\hspace{1cm}}$ | $2 + 7 = \underline{\hspace{1cm}}$ | $3 + \underline{\hspace{1cm}} = 10$ | $2 + \underline{\hspace{1cm}} = 9$ |
| $4 + 4 = \underline{\hspace{1cm}}$ | $5 + 3 = \underline{\hspace{1cm}}$ | $3 + \underline{\hspace{1cm}} = 8$ | $2 + \underline{\hspace{1cm}} = 7$ |

7. Draw the missing marbles to match the addition sentence.

| | |
|---|---|
| a. $6 + 1 + \underline{\hspace{1cm}} = 10$ | b. $1 + 4 + \underline{\hspace{1cm}} = 8$ |

8. Write the fact families.

a. Numbers: 9, 5, 4

$\underline{\hspace{1cm}} + \underline{\hspace{1cm}} = \underline{\hspace{1cm}}$

$\underline{\hspace{1cm}} + \underline{\hspace{1cm}} = \underline{\hspace{1cm}}$

$\underline{\hspace{1cm}} - \underline{\hspace{1cm}} = \underline{\hspace{1cm}}$

$\underline{\hspace{1cm}} - \underline{\hspace{1cm}} = \underline{\hspace{1cm}}$

b. Numbers: 10, 2, 8

$\underline{\hspace{1cm}} + \underline{\hspace{1cm}} = \underline{\hspace{1cm}}$

$\underline{\hspace{1cm}} + \underline{\hspace{1cm}} = \underline{\hspace{1cm}}$

$\underline{\hspace{1cm}} - \underline{\hspace{1cm}} = \underline{\hspace{1cm}}$

$\underline{\hspace{1cm}} - \underline{\hspace{1cm}} = \underline{\hspace{1cm}}$

9. Draw marbles for the child that has none.

| | |
|---|---|
| [ ] Jane | [marbles] Louis |
| [marbles] George | [ ] Henry |
| **a.** Jane has 2 more than George. | **b.** Louis has 4 more than Henry. |
| [ ] Roger | [ ] John |
| [marbles] Peter | [marbles] Gwen |
| **c.** Roger has 2 fewer than Peter. | **d.** Gwen has 3 fewer than John. |

10. Solve.

**a.** Three children were playing. Then, five more children came to play.
Then, one child left. Now how many children are playing?

**b.** Judy has 3 marbles and Amanda has 7.
How many marbles do the girls have together?

How many more marbles does Amanda have than Judy?

**c.** Kyle has 10 toy trucks. Some are blue and seven are black.
How many are blue?

**d.** Leah has $6. She wants to buy a book for $9.
How much more money does she need?

**e.** Matthew has 10 socks and he cannot find any of them!
Then, he found three socks under the bed and five in the closet.
How many socks did Matthew find?

How many are still missing?

# Cumulative Review, Grade 1, Chapters 1-3

1. Break the numbers into tens and ones.

| a. 22 = _____ + _____ | b. 64 = _____ + _____ | c. 95 = _____ + _____ |
|---|---|---|

2. Compare. Write < , > or = .

| a. 2 + 3 ☐ 5 + 1 | c. 8 − 2 ☐ 4 | e. 6 ☐ 4 + 2 |
|---|---|---|
| b. 6 − 4 ☐ 8 + 2 | d. 7 − 4 ☐ 5 | f. 8 ☐ 9 − 1 |

3. Write the fact families.

| a. 3 + 7 = _____ | b. 6 + _____ = 9 |
|---|---|
| _____ + _____ = _____ | _____ + _____ = _____ |
| _____ − _____ = _____ | _____ − _____ = _____ |
| _____ − _____ = _____ | _____ − _____ = _____ |

4. Skip-count by tens.

a. 4, 14, _____, _____, _____, _____, _____, _____

b. _____, _____, _____, 68, 78, _____, _____, _____

5. a. Skip-count by fives starting at 45, and colour all those numbers yellow.

b. Colour all the even numbers blue.

Which numbers end up green?

| 41 | 42 | 43 | 44 | 45 | 46 | 47 | 48 | 49 | 50 |
|---|---|---|---|---|---|---|---|---|---|
| 51 | 52 | 53 | 54 | 55 | 56 | 57 | 58 | 59 | 60 |
| 61 | 62 | 63 | 64 | 65 | 66 | 67 | 68 | 69 | 70 |

6. Name and write the numbers.

   **a.** 1 ten 1 one _____

   **b.** 1 ten 7 ones _____

   **c.** 1 ten 5 ones _____

   **d.** 1 ten 3 ones _____

7. Find the difference between the numbers. "Travel" on the number line!

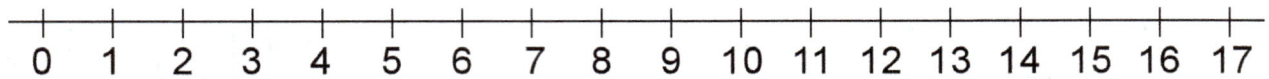

| 0 | 1 | 2 | 3 | 4 | 5 | 6 | 7 | 8 | 9 | 10 | 11 | 12 | 13 | 14 | 15 | 16 | 17 |

| From | 2 | 11 | 9 | 14 | 6 | 12 | 6 | 10 |
|---|---|---|---|---|---|---|---|---|
| To | 10 | 7 | 9 | 7 | 6 | 5 | 12 | 15 |
| Difference | | | | | | | | |

8. Write < , > or = .

| **a.** 82 ☐ 29 | **c.** 70 + 4 ☐ 7 + 40 | **e.** 60 + 7 ☐ 70 + 5 |
|---|---|---|
| **b.** 75 ☐ 67 | **d.** 20 + 8 ☐ 2 + 80 | **f.** 2 + 90 ☐ 9 + 50 |

9. Solve.

**a.** Some children needed 10 players for a game. They already had 2 boys and 2 girls. How many more children do they need for their game?

**b.** A herd has 10 brown horses, 20 white horses, and 10 speckled ones. How many horses are there in the herd?

How many more white horses are there than brown ones?

# Cumulative Review, Grade 1, Chapters 1-4

1. Pick a number so the comparison is true.

| 3  4  5<br><br>$2 + \_\_\_\_ < 6$ | 4  5  6<br><br>$1 + \_\_\_\_ > 6$ | 3  4  5<br><br>$4 + \_\_\_\_ < 8$ |
|---|---|---|

2. Add.

a. $0 + 4 + 2 = \_\_\_\_$      b. $7 + 1 + 1 = \_\_\_\_$      c. $2 + 5 + 3 = \_\_\_\_$

3. Fill in the numbers and name them.

| a. _____ | b. _____ | c. _____ |
|---|---|---|
| \_\_\_\_ + \_\_\_\_ | \_\_\_\_ + \_\_\_\_ | \_\_\_\_ + \_\_\_\_ |
| tens    ones | tens    ones | tens    ones |

4. The numbers are broken into tens and ones. Fill in the missing parts.

| a. $40 + \_\_\_\_ = 48$ | b. $\_\_\_\_ + \_\_\_\_ = 62$ | c. $50 + 5 = \_\_\_\_$ |
|---|---|---|

5. What number is…

| a. | b. | c. |
|---|---|---|
| one more than 16 \_\_\_\_ | two more than 11 \_\_\_\_ | ten more than 12 \_\_\_\_ |
| one less than 29 \_\_\_\_ | two less than 67 \_\_\_\_ | ten less than 30 \_\_\_\_ |
| one less than 40 \_\_\_\_ | two more than 59 \_\_\_\_ | ten more than 87 \_\_\_\_ |

6. Count. You can also do this orally with your teacher.

96, 97, _____ , _____ , _____ , _____ , _____ ,

_____ , _____ , _____ , _____ , _____ , _____

7. Draw tally marks for these numbers.

| a. 9 | b. 11 |
|------|-------|
| c. 27 | d. 32 |

8. Some children counted their stuffed animals.

   a. How many does Alice have?

   b. How many does Aaron have?

   c. How many does Maria have?

   d. Alice gave 10 stuffed animals to Aaron. Now how many does Alice have?

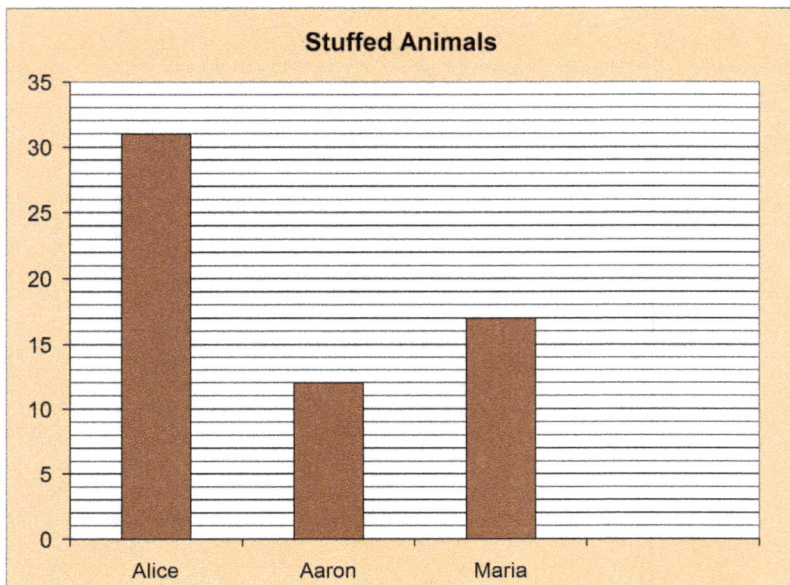

How many does Aaron have now?

In the empty space on the graph you can draw a bar for how many stuffed animals you have.

9. Find the "mystery numbers"! You will need to think logically.

| a. This number has three more tens than 50 has, and the same amount of ones as 13. | b. This number has seven less ones than 29, and six more tens than 17. |
|------|------|

# Cumulative Review, Grade 1, Chapters 1 - 5

1. Add.

| | | | |
|---|---|---|---|
| a. $1 + 4 =$ _____ | b. $5 + 2 =$ _____ | c. $3 + 6 =$ _____ | d. $7 + 3 =$ _____ |

2. Subtract.

| a. | b. | c. | d. |
|---|---|---|---|
| $5 - 2 =$ _____ | $9 - 4 =$ _____ | $7 - 3 =$ _____ | $10 - 8 =$ _____ |

| e. | f. | g. | h. |
|---|---|---|---|
| $6 - 2 =$ _____ | $10 - 7 =$ _____ | $7 - 7 =$ _____ | $9 - 5 =$ _____ |

3. Write the names of the numbers with whole tens.

two tens      _____

three tens      _____

eight tens      _____

five tens      _____

4. Put the numbers in order from the smallest to the largest.

| a. 58, 17, 36 | b. 23, 63, 36 | c. 48, 84, 44 |
|---|---|---|
| _____ < _____ < _____ | _____ < _____ < _____ | _____ < _____ < _____ |

5. Solve the missing numbers.

| | | |
|---|---|---|
| a. $8 - \boxed{\phantom{00}} = 5$ | b. $\boxed{\phantom{00}} - 2 = 4$ | c. $\boxed{\phantom{00}} - 1 = 9$ |
| d. $\boxed{\phantom{00}} + 3 = 9$ | e. $2 + \boxed{\phantom{00}} = 9$ | f. $9 - \boxed{\phantom{00}} = 9$ |

6. Draw the hour hand on the clocks. Then write the time that the clock will show a half-hour later.

| | a. two o'clock | b. ten o'clock | c. half-past six | d. half-past eight |
|---|---|---|---|---|
| 1/2 hour later → | _____ | _____ | _____ | _____ |

7. Fill in either AM or PM.

| | |
|---|---|
| **a.** You woke up. It was 7 _____. | **b.** John plays in the afternoon at 3 _____. |
| **c.** Jack is asleep. It is dark. It is 1 _____. | **d.** It is time for lunch. It is 1 _____. |

8. Compare. Write < , > or = .

| | | |
|---|---|---|
| **a.** 62 ☐ 3 + 60 | **c.** 10 – 2 ☐ 7 | **e.** 7 ☐ 9 – 2 |
| **b.** 54 ☐ 42 + 10 | **d.** 6 – 5 ☐ 0 | **f.** 45 ☐ 65 – 10 |

9. Solve.

**a.** A baby put some of his ten blocks into a bucket. He had four blocks left on the floor. So, how many blocks did he put into the bucket?

**b.** Tanya has $10. She gets another $4 from her mom.

Now how much does she have?

How much more does she need to buy a watch for $20?

# Cumulative Review, Grade 1, Chapters 1 - 6

1. Find the missing numbers.

| | | | |
|---|---|---|---|
| **a.** $7 +$ _____ $= 7$ | **b.** _____ $+ 6 = 9$ | **c.** $4 +$ _____ $= 9$ | **d.** $8 +$ _____ $= 10$ |

2. Cross out the problem if you cannot take away that many.

$4 - 6$ $\qquad$ $2 - 0$ $\qquad$ $6 - 9$ $\qquad$ $8 - 4$

3. Write these numbers with words. You can also do this orally with your teacher.

**a.** 2 tens 9 ones = _____

**b.** 9 tens 1 one = _____

**c.** 1 ten 5 ones = _____

**d.** 5 tens 7 ones = _____

4. Write the time for a half-hour later. Use numbers.

| Now it is: | **a.** 1:00 | **b.** 11:30 | **c.** 9:00 | **d.** 6:30 | **e.** 4:00 |
|---|---|---|---|---|---|
| a half-hour later, it is: | ___ : ___ | ___ : ___ | ___ : ___ | ___ : ___ | ___ : ___ |

5. Draw the different shapes that you have learned, and label them.

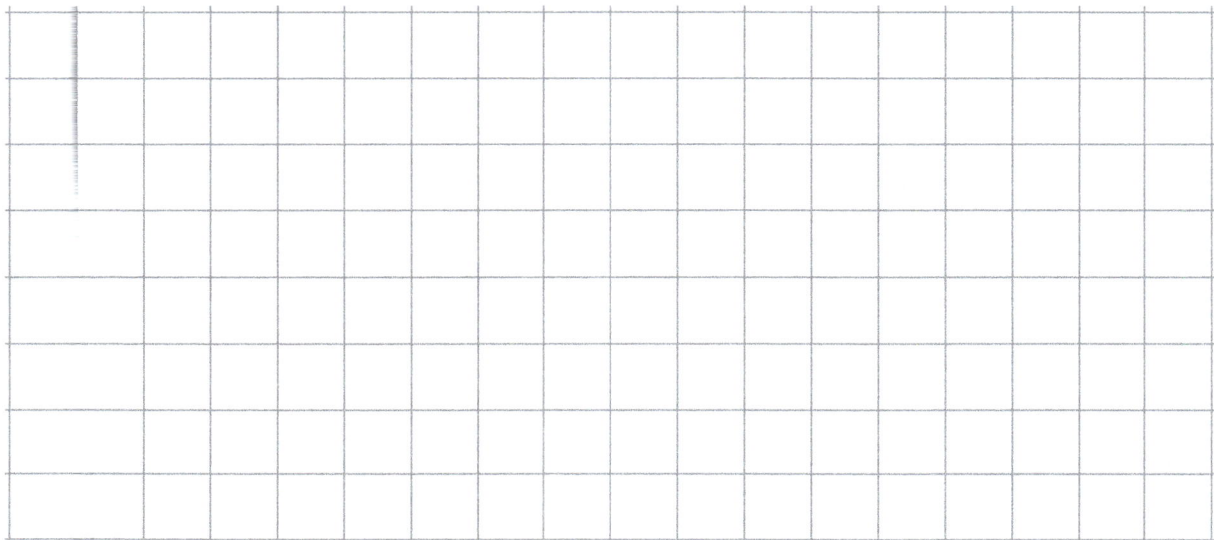

6. Colour the circles yellow,
   the triangles blue,
   the squares pink,
   the rectangles green
   and the rest of the
   shapes red.

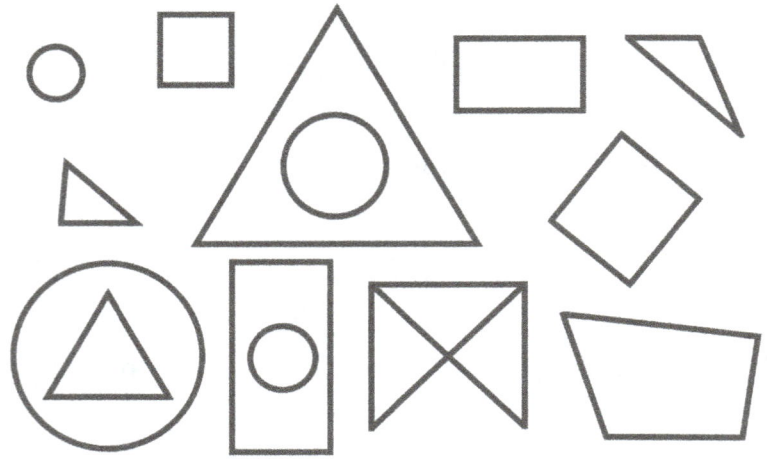

7. Solve.

| | | |
|---|---|---|
| $9 - \underline{\phantom{xx}} = 4$ | $3 + \underline{\phantom{xx}} = 7$ | $10 - \underline{\phantom{xx}} = 8$ |
| $3 + \underline{\phantom{xx}} = 4$ | $8 - \underline{\phantom{xx}} = 7$ | $3 + \underline{\phantom{xx}} = 8$ |
| $10 - \underline{\phantom{xx}} = 4$ | $0 + \underline{\phantom{xx}} = 7$ | $9 - \underline{\phantom{xx}} = 8$ |
| $2 + \underline{\phantom{xx}} = 4$ | $9 - \underline{\phantom{xx}} = 7$ | $2 + \underline{\phantom{xx}} = 8$ |
| $9 - \underline{\phantom{xx}} = 6$ | $3 + \underline{\phantom{xx}} = 5$ | $10 - \underline{\phantom{xx}} = 3$ |
| $3 + \underline{\phantom{xx}} = 6$ | $10 - \underline{\phantom{xx}} = 5$ | $3 + \underline{\phantom{xx}} = 3$ |
| $10 - \underline{\phantom{xx}} = 6$ | $0 + \underline{\phantom{xx}} = 5$ | $7 - \underline{\phantom{xx}} = 3$ |
| $2 + \underline{\phantom{xx}} = 6$ | $9 - \underline{\phantom{xx}} = 5$ | $2 + \underline{\phantom{xx}} = 3$ |

8. Write < or > between the numbers to compare them.

a. $30 \boxed{<} 38$    b. $87 \boxed{\phantom{x}} 85$    c. $69 \boxed{\phantom{x}} 96$    d. $58 \boxed{\phantom{x}} 56$

e. $60 \boxed{\phantom{x}} 48$    f. $43 \boxed{\phantom{x}} 95$    g. $49 \boxed{\phantom{x}} 94$    h. $22 \boxed{\phantom{x}} 32$

# Cumulative Review, Grade 1, Chapters 1 - 7

1. Find the missing numbers.

| a. | b. | c. | d. |
|---|---|---|---|
| 2 + _____ = 4 | 2 + _____ = 7 | 6 – _____ = 6 | 3 – 1 = _____ |
| 5 + _____ = 9 | 0 + _____ = 5 | 8 – _____ = 4 | 10 – 3 = _____ |

2. Find the "mystery numbers"! You will need to think logically.

| a. This number has one more ten than 20 has, and the same amount of ones as 63. | b. This number has two less ones than 88, and six more tens than 24. |
|---|---|

3. Count by tens.

16, 26, _____, _____, _____, _____, _____, _____

4. Find how much the two items cost together.

| a. A tropical fish, $12, and a parakeet, $65 | b. A backpack for $76, and a torch, $23 |
|---|---|
| Together they cost | Together they cost |
| $ _____ . | $ _____ . |

5. Draw the hands to show the time.

| a. 5 o'clock | b. half past seven | c. 11 o'clock | d. half past two |
|---|---|---|---|

6. Fill in the blanks.

a. _____ has four sides the same length.

b. _____ has three sides and three vertices.

7. Add using the "nine trick" and the "eight trick".

| a. 9 + 8 = _____ | b. 9 + 3 = _____ | c. 9 + 5 = _____ |
| 8 + 8 = _____ | 8 + 4 = _____ | 7 + 8 = _____ |

8. Subtract and add whole tens.

| a. 25 + 10 = _____ | b. 90 − 30 = _____ | c. 92 − 10 = _____ |
| 60 + 20 = _____ | 100 − 70 = _____ | 64 − 10 = _____ |

9. Divide these shapes by drawing straight lines from dot to dot. Then colour.

| a. | b. | c. | d. |
| Colour one quarter. | Colour two quarters. | Colour two halves. | Colour three fourths. |

10. Which is more, two quarters or one half?
Colouring the parts in the pictures may help.

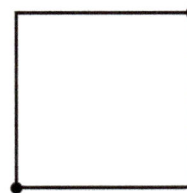

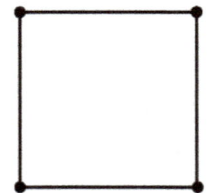

11. Name the basic shape. Is it a cylinder, a cube, a box, or a ball?

a.          b.          c.          d.

# Cumulative Review, Grade 1, Chapters 1 - 8

1. Find the missing numbers.

| a. | b. | c. | d. |
|---|---|---|---|
| _____ $- 1 = 9$ | $8 + 9 =$ _____ | $25 - 10 =$ _____ | $52 + 7 =$ _____ |
| _____ $- 2 = 6$ | $7 + 8 =$ _____ | $38 - 10 =$ _____ | $35 + 3 =$ _____ |
| _____ $- 3 = 4$ | $5 + 6 =$ _____ | $100 - 10 =$ _____ | $26 + 2 =$ _____ |

2. Draw the hour hand.

| | | |
|---|---|---|
| **a.** half past one | **b.** half past five | **c.** half past ten |

3. You can either cut out the shapes to do the exercises, or if you can, imagine putting the shapes together; then sketch (draw) the shapes that illustrate the answers.

**a.** Put together the #5 triangles so that you get a rectangle.

**b.** Now place them together so that you get a different four-sided shape (not a rectangle).

**c.** Use some of the #1 (yellow) triangles to form the purple shape (#6)

**d.** Now use a #1 triangle and the purple shape. What shapes can you make with these two?

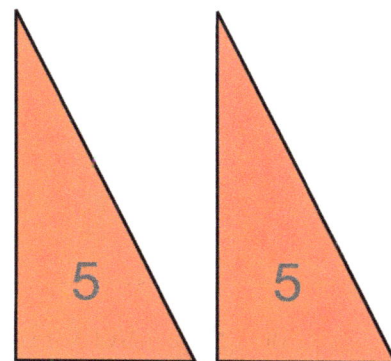

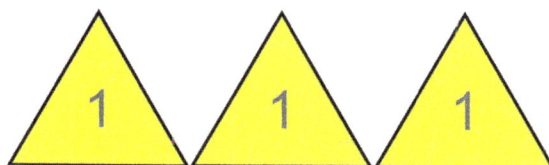

4. First subtract to 10. Then subtract the rest.

| a. 15 − 7 | b. 14 − 9 | c. 16 − 8 |
|---|---|---|
| 15 − ___ − ___ | 14 − ___ − ___ | 16 − ___ − ___ |
| = ___ | = ___ | = ___ |

5. Add and subtract.

a.
$$\begin{array}{r} 2\ 5 \\ +\ 3\ 0 \\ \hline \end{array}$$

b.
$$\begin{array}{r} 5\ 1 \\ +\ 3\ 4 \\ \hline \end{array}$$

c
$$\begin{array}{r} 7\ 8 \\ -\ 1\ 5 \\ \hline \end{array}$$

d.
$$\begin{array}{r} 8\ 6 \\ -\ 4\ 4 \\ \hline \end{array}$$

6. Draw circles with "5", "10" and "25" to make these amounts of money.
Use the least number of coins possible.

| a. 35¢ | b. 15¢ | c. 70¢ |
|---|---|---|
| d. 80¢ | e. 100¢ | f. 45¢ |

7. You bought an item. How much money do you have left?

| a. You had: | You bought a banana for 60¢.<br><br>How much is left?<br><br>_____ ¢ | b. You had: | You bought an eraser for 85¢.<br><br>How much is left?<br><br>_____ ¢ |
|---|---|---|---|